T0321618

CLARENCE EDWARD DUTTON

Clarence Edward Dutton. Courtesy of Special Collections,
J. Willard Marriott Library, University of Utah.

Clarence Edward Dutton

An Appraisal

by

WALLACE STEGNER

Foreword by

PHILIP L. FRADKIN

THE UNIVERSITY OF UTAH PRESS

Salt Lake City

Foreword © 2006 by Philip L. Fradkin

Clarence Edward Dutton: An Appraisal by Wallace E. Stegner originally published by the University of Utah, ca. 1936. Reprinted by permission of Brandt and Hochman, Literary Agents, Inc.

This publication contains a facsimile of the original, from Special Collections, J. Willard Marriott Library, University of Utah.

 The Defiance House Man colophon is a registered trademark of the University of Utah Press. It is based upon a four-foot-tall, Ancient Puebloan pictograph (late PIII) near Glen Canyon, Utah.

10 09 08 07 06   5 4 3 2 1

LIBRARY OF CONGRESS CATALOGING-IN-PUBLICATION DATA

Stegner, Wallace Earle, 1909-1993.
   Clarence Edward Dutton : an appraisal / by Wallace E. Stegner ; foreword by Philip L. Fradkin.
      p. cm.
   "Originally published by the University of Utah, ca. 1936"—T.p. verso.
   Includes bibliographical references.
   ISBN-13: 978-0-87480-865-0 (alk. paper)
   ISBN-10: 0-87480-865-0 (alk. paper)
   1. Dutton, Clarence E. (Clarence Edward), 1841-1912. 2. Geologists—United States—Biography.   I. Title.
   QE22.D85S7 2006
   551.092—dc22
   [B]
                                    2006012961

CONTENTS

FOREWORD

Philip L. Fradkin

A finger smashed in a car door, a missed geology examination at the University of Utah, and a special make-up assignment led Wallace Stegner to Clarence E. Dutton, thence to John Wesley Powell, and finally in 1954 to publication of what is arguably the single best nonfiction book dealing with the American West. *Beyond the Hundredth Meridian* remains in print more than a half-century after its initial appearance. Not many books have survived that length of time and thrived, especially those with competing works on the same subject. This combination biography, history, and environmental primer, written with the flair and the technical skill of a novelist who could masterfully evoke scenes and sustain a gripping factual narrative, sprang from "Clarence Edward Dutton: An Appraisal" (University of Utah, 1936). Seldom has such a classic book had such a humble beginning.

Stegner was a young English instructor at the University of Utah when he produced the Dutton essay. He was ambitious and desperate for recognition, a raise (he was earning $1,700 a year), and steady employment in the Depression years. The essay contains hints, in terms of style and content, of what Stegner would eventually produce. "Dutton" was Stegner's first published work of nonfiction, and it is fair to say that it led him, in conjunction with Bernard DeVoto's prodding, to the subject of conservation.

Born in Iowa, Stegner moved with his family at an early age to the Puget Sound area, where his first memory was of a Seattle orphanage. He was subsequently molded by the harsh realities and outdoor life of the Saskatchewan prairie and by the deserts, mountains, rivers, and canyonlands of Utah. To many westerners transience was no stranger, and the same can be said for Stegner. But he absorbed those experiences and gave them meaning. In fact, he became, in my opinion, the greatest writer of fiction and nonfiction the West has produced, as well as an active conservationist at the local and national levels and the teacher of many famed writers at Stanford University, where he founded and directed the creative writing program for many years.

For all these reasons and more, this initial exercise deserves to be hauled out of a few obscure archives, reproduced faithfully, and made available to an extensive audience. "Clarence Edward Dutton: An Appraisal" was Stegner's nonfiction genesis. What resulted from this obscure birthing, namely *Beyond the Hundredth Meridian* and its offspring, written by Stegner and others who he had influenced, was a powerful evocation and accurate description of an arid land. Aridity is the ultimate reality of the American West, a reality that still too few are willing to acknowledge. We ignore Stegner at our peril.

While on camping trips as a Boy Scout in Salt Lake City, Stegner explored the Granddaddy Lakes in the Uinta Mountains, Split Mountain on the Green River, and the canyonlands of southern Utah. He traveled with his scoutmaster and his son in 1924 through Zion, Grand Canyon, and Bryce national parks, the latter then known as Utah National Park. The Stegner family had a cabin at the nine-thousand-foot level of Fish Lake in southern Utah. There he hiked, swam, hunted, and fished among the aspen trees on the cool heights and viewed the heat-soaked canyonlands in the distance. Stegner would soon have an opportunity to deepen his historical and geologic understanding of these youthful experiences.

A friend smashed his thumb in a car door at the university, and Stegner accompanied him to the hospital when he was supposed to be taking a final geology exam from Frederick J. Pack. Instead of a make-up exam, Professor Pack gave him a copy of Dutton's *Report on the Geology of the High Plateaus of Utah* to read. It was a book he could relate to from first-hand experience. "I think that's the first time that my active, existential life happened to cross the track of history," Stegner said years later.[1] Once bitten by the history bug, he was infected for life.

From Utah, Stegner went to the University of Iowa, where the written requirements for advanced degrees had been relaxed. He wrote two short stories for his master's degree in creative writing, but he was advised to find a more lucrative field in which to specialize for his Ph.D.[2] He chose American literature, not a particularly promising subject because it was considered by the dominant Europeanists at the time as somewhat of an oxymoron. Norman Foerster, the director of the School of Letters, encouraged good writing. Foerster allowed graduate students to write innovative, scholarly articles for their doctors' thesis rather than dry, academic dissertations.

It was Foerster, the author of *Nature in American Literature,* who suggested that Stegner pick a western naturalist for his thesis. Stegner chose the literary scientist Clarence Edward Dutton, with whom he was already acquainted. In later years Stegner was dismissive of the effort. "As a matter of fact," he said, "I did a very bad Ph.D dissertation on Dutton, a most unlikely subject for a Ph.D dissertation in English because he was a geologist examining the southern Utah plateaus and the Grand Canyon."[3]

Stegner arrived back on the campus of the University of Utah in September 1934 as a newly-minted doctor of American literature, a just-married husband, and a lowly English instructor. While his undergraduate years were among the highlights of his life, the three years he taught at the university were among the low points. He made every effort to advance, but was continually rebuffed.

One such effort was to get a condensed version of his thesis in print, a traditional approach to the publish or perish syndrome that is practiced to this day. From the very start of his writing career to the end, the thrifty Stegner was expert at recycling his works.[4] There are three versions of the thesis: the original thesis titled "Clarence Edward Dutton, Geologist and Man of Letters" (1935); "Clarence Edward Dutton: An Appraisal" (1936); and "C.E. Dutton, Explorer, Geologist, Nature Writer," *Scientific Monthly,* July 1937.[5]

"An Appraisal," Stegner said, was "a sort of abstract of the thesis. The University of Utah Press made itself available for publication by local faculty. I dug it out; it was after all on a local subject." Asked by an interviewer if he had been pressured to publish, Stegner replied, "Well, you bet. Those were the years when there was no promotion whatever in any college."[6]

The university newspaper, the *Utah Chronicle*, took notice of Stegner's publication two years later. It quoted the English instructor as stating that Dutton's work was "the finest thing from the standpoint of literary merit in the whole history of American Geology."[7]

John Wesley Powell, director of the U.S. Geological Survey and Dutton's boss, makes a brief appearance in the essay as "a great organizer with 'extraordinarily fertile ideas.'" But the seventeen pages of text and three pages of bibliographic information belong almost entirely to Dutton. Powell would get his due and be elevated from relative obscurity to his rightful place in American history in the much longer work that built on the foundation of the Dutton essay.

From the start, Stegner was mindful of structure. The essay is divided into three parts: Dutton's character, his early background and later professional achievements, and his literary style. The structure was stiff. After all, it was an academic exercise boiled down for an academic press for the purpose of acquiring academic credits. But Stegner was pressing boundaries at the time (as he would later by mixing fiction and nonfiction techniques in works of one or the other genre) by judging a scientific career on its literary merits.

The work was partially self serving, an emphatic no-no for thesis writers who are supposed to first and foremost serve their disciplines. When he wrote about Dutton, Stegner was also describing his own writerly qualities, what he admired, and what he hoped to achieve. The Dutton essay was, in effect, a mirror in a literary changing room where Stegner was trying on different styles. By examining the work of another author, who just happened to be a geologist, Stegner was, in his midtwenties, learning the craft of writing.

We can learn a great deal about Stegner by observing what he wrote about Dutton. To Stegner, "Dutton was the farthest thing from a narrow specialist."[8] The description also fit Stegner, and it was one of the reasons why he was attracted to both Dutton and Powell. Dutton was one of the "pioneers" in a new land. He associated with such "men of large caliber" as Henry Adams, Grove Karl Gilbert, Ferdinand Hayden, Clarence King, and Powell, all of whom would make multiple appearances in *Beyond the Hundredth Meridian*. King can also be found in the Pulitzer Prize–winning novel *Angle of Repose*. These were men of "concrete achievements." That his family lacked such achievements (his father was a bootlegger) was both a personal and professional motivating factor for Stegner.

Underneath Dutton's natural reserve was a "teeming mind." He read voraciously, had a "powerful memory," wrote rapidly, and spoke "with remarkable fluency." Stegner admired "the orderliness and concentration of Dutton's mind" and the fact that "he allows himself to show through his books, which is not often the case. . . ." All of these qualities would become Stegner characteristics, including "the clear, pointed sentences [that] emerged, as it were, from a moving cloud of cigar smoke." Stegner also smoked a cigar while he wrote.

The East-versus-West theme that emerges in much of Stegner's work, particularly in his two masterpieces, *Beyond the Hundredth Meridian* and *Angle of Repose,* makes its first appearance here. Citing the humor and alliteration sprinkled through Dutton's public lectures, especially the one on the Hawaiian Islands, Stegner particularly liked "the comparison of the full-blooded and buxom girls of Hawaii with their 'pale pious and pulmonary sisters of the effete East.'" Stegner admired Dutton's "gusto" and his sense of humor, which reminded him of Mark Twain—a favorite of both men.

There were some superficial similarities in their early backgrounds. Both were two years ahead of their classmates and both discovered writing at their separate universities. Stegner was one of Vardis Fisher's students and edited the campus literary magazine; Dutton won the literary prize at Yale University. The winning essay,

Stegner thought, showed "some youthfulness of thought and a religiosity that Dutton later lost." It also demonstrated "wide reading, compact and close organization, and accurate and flexible phrasing."

Stegner could hardly wait to discuss Dutton's literary style, references to it having appeared in the first two sections of the essay.[9] Dutton's use of words had the greatest impact on the budding writer, and Stegner grew rhapsodic in his praise. The Dutton of Stegner's twenties emerges flawless and heroic. A few academics would later take issue with Stegner's portrait of Powell, believing that it was too uncritical. In comparison to Dutton, however, his treatment of Powell seems subdued.[10]

First, there were some basic lessons to learn from Dutton. Know your subject: be there, see it, experience it, absorb it, feel it, and understand it. "He knew the canyon and the region surrounding it as well as any man of his time," Stegner wrote, "not excepting Major Powell, and he knew not only its appearance, but its meaning."

Facts, in this case geologic, should anchor description and both should serve a narrative purpose. "Literary landscape painting in itself is an inert and lifeless thing unless bolstered by narrative or movement of one sort or another. But Dutton goes one step further, and not only gives his landscapes vigor and life by a narrative frame, but also adds a great deal to their stamina by subordinating them to the larger purpose of geological interpretation." Stegner, like the geologist, employed long series of words or phrases linked by commas. For both men serial phrases were a descriptive tic.

Next in Stegner's writing canon came clarity of expression, followed by an accurate eye for color and form that should be true to what it observes, not what it wants to see. Scenery should not be preconceived, meaning that the canyonlands of the Southwest should not be mistaken for the Alps of Europe. Stegner believed that "within his range [Dutton] wrote as well as any American has ever done." That Dutton was given to excesses of description—overwriting, in more modern terms—was not apparent to young Stegner. Dutton's "modulated" enthusiasm, he believed, was not "the wild ecstatic thing" that most tourists applied to the West.

But the prose that mainly charmed was also overheated at times, as Stegner realized nearly twenty years later in *Beyond the Hundredth Meridian* when he described Dutton's verbal landscape painting as casting "a romantic and exciting aura" over the realities of the West. Stegner compared Dutton's verbal descriptions to the Turneresque school of painting as it was practiced by Thomas Moran, who accompanied the geologist.[11] By the mid-twentieth century, the romanticism that was a remnant of the previous century had been tempered by Stegner's familial responsibilities, middle age, his passage through depression, world war, the cold war, work at four universities, numerous changes of address, and acceptances and rejections of many manuscripts.

The best evidence of Stegner's growth is contained in a comparison of what Stegner thought in 1935 was one of the most outstanding descriptive passages in *Tertiary History* (see pages eighteen and nineteen of the essay) and the closest he came to writing a Duttonesque passage in *Beyond the Hundredth Meridian* in 1954.

Stegner is describing Powell's first trip through the Glen Canyon portion of the Colorado River in 1869, a journey Stegner partially duplicated in 1947 in order to gather material for the book:

> Through most of its course the canyoned Green and Colorado, though impressive beyond description, awesome and colorful and bizarre, is scenically disturbing, a trouble to the mind. It works on the nerves, there is no repose in it, nothing that is soft. The water-roar emphasizes what the walls begin: a restlessness and excitement and irritability. But Glen Canyon, into which they now floated and which they first called Monument Canyon from the domes and "baldheads" crowning its low walls, is completely different. As beautiful as any of the canyons, it is almost absolutely serene, an interlude for a pastoral flute.[12]

Stegner had been there. He knew the place and applied the novelist's palette to it. Not everything is perfect or beautiful. There is movement in the overall narrative and a rhythm in the individual sentences that matches the descriptions. For instance, tension pervades the "no repose" sentence. Colors and forms and textures emerge just subtly enough to encourage the imagination and not bludgeon the reader. This land of bare rocks cannot be mistaken for any other exotic place. Stegner's concluding words, placed in the most powerful position in a sentence or a paragraph, compel the reader forward with a musical resonance that fits the advancing scene.

Gone is the florid prose of the nineteenth century. For twentieth-century sensibilities, and those of the present century, Stegner fits much better. He had shed Dutton, as a graduating pupil needs to be rid of a teacher, and had moved forward on his own terms and in his own voice, as a mature writer must do. In 1977 Stegner still found Dutton's *Tertiary History* "astonishingly fresh after nearly a hundred years" and deserving of a general audience.[13] The same can now be said of Stegner's major works on the eve of the centennial of his birth in 1909.

Stegner believed in the relevancy of history, the fact that history lives within all of us both on a culturally collective and an individual basis. "The past controls us a whole lot more than you might want to be controlled," he said.[14] We need to know where Wallace Stegner, the nonfiction writer, derived his literary skills in order to understand how he arrived at what he eventually became. This reprinted essay for the first time gives a wider readership the opportunity to make that determination.

NOTES

1. Wallace Stegner, "Literary by Accident," *Utah Libraries* (Fall 1975): 12. There are a number of discrepancies in Stegner's accounts, and I have chosen what seems most accurate. In this article written by Stegner, the author said his friend's finger was injured in Stegner's freshman year, and he had to take him to the hospital "where it was patched up and splints put on it." Seven years later Stegner told an interviewer that it was "my" thumb that was smashed in the car door in his sophomore year. Wallace Stegner, interview by Ann Lage, 1982, Regional Oral History Office, Bancroft Library, University of California at Berkeley, 3. University records show that Stegner took Pack's two-part course in his junior year. Pack was

Deseret Professor of Geology and chair of the Geology Department for thirty-one years. He taught a general course in geology "from a cultural viewpoint." Stegner took the course in the fall and winter quarters of 1928–29 and received a B in both classes. (The confusion in Stegner's mind many years later may have come from the fact that he had to drop out of two of the three quarters in the academic year of 1927–28, when he would have been a junior, in order to earn money. Thus his junior year was delayed to 1928–29. He graduted in 1930, one year later than he normally would have.) Kirk Baddley, personal communication, University of Utah Archives.

2. This program soon became known as the Iowa Writers' Workshop.

3. Stegner, "Literary by Accident," 12. *Tertiary History* was a rare book then and now and the "penniless graduate student" (Stegner's words) could not afford to purchase a copy. So he transcribed the entire text by typewriter. "But as soon as I could afford it, I paid the price, because it was a book I wanted to own." Wallace Stegner, "Introduction," in Clarence E. Dutton, *Tertiary History of the Grand Cañon District* (Santa Barbara and Salt Lake City: Peregrine Smith, 1977), ix.

4. One such example is the short story "Genesis" (1959), incorporated into a nonfiction book *Wolf Willow* (1962), and then chosen for two collections of short stories, *Collected Stories of Wallace Stegner* (1990) and *Marking the Sparrow's Fall* (1998), edited by Page Stegner.

5. There were so many versions that Dale L. Morgan, seeking bibliographic information for his book on the Great Salt Lake, wrote Stegner and asked if he could sort them out for him. Dale L. Morgan to Wallace Stegner, April 30, 1946, Special Collections, University of Utah. Stegner again wrote about the geologist in the introduction to the 1977 Peregrine Smith reprint of Dutton's *Tertiary History of the Grand Cañon District*. That introduction was condensed in an issue of *American West* the next year and has been preserved in the University of Arizona Press's 2001 edition of *Tertiary History*, another example of publication recycling.

6. Wallace Stegner and Richard W. Etulain, *Conversations with Wallace Stegner on Western History and Literature* (Salt Lake City: The University of Utah Press, 1990), 26–27, 37.

7. *Utah Chronicle*, January 7, 1937.

8. Stegner, "Introduction," viii; Wallace Stegner, "The Scientist as Artist: Clarence E. Dutton and the *Tertiary History of the Grand Cañon District*," *American West* (May/June 1978): 19.

9. That the geologist possessed a literary style and had employed it for a purpose was acknowledged by Dutton in his preface to *Tertiary History*. "I have in many places departed from the severe ascetic style which has become conventional in scientific monographs." He had done this, he said, "to exalt the mind sufficiently to comprehend the sublimity of the subjects." Dutton, *Tertiary History*, xvi.

10. The criticisms come from historians who have been certified by advanced degrees in history. Stegner had no such degree. Thomas G. Manning, *"Beyond the Hundredth Meridian," The American Historical Review* (January 1955): 389–90; Donald Worster, *A River Running West: The Life of John Wesley Powell* (New York: Oxford University Press, 2001), xii; Donald Worster, personal communication; Richard White, "Nature or Justice," *The New Republic* (June 11, 2001): 47–52. White, a professor of history at Stanford University, wrote of *Beyond the Hundredth Meridian,* "Stegner was a skilled writer, but he was not much of a historian." Stegner's academic career, especially in its early stages, was not helped by the fact that he wrote novels that sold relatively well.

11. I have not cited the quotes extracted from "An Appraisal," as they are numerous and can easily be located by the reader in the accompanying text. For other quotations, see Wallace Stegner, *Beyond the Hundredth Meridian: John Wesley Powell and the Second Opening of the West* (New York: Penguin Books, 1992), 165.

12. Stegner, *Beyond the Hundredth Meridian,* 88.

13. Stegner, "Introduction," viii.

14. Forrest G. Robinson and Margaret G. Robinson, "Wallace Stegner, an Interview," *Quarry* 74 (1974): 79.

Clarence Edward Dutton

An Appraisal

BY

Wallace E. Stegner
University of Utah

Published by
UNIVERSITY OF UTAH
Salt Lake City

THE UNIVERSITY PRESS
UNIVERSITY OF UTAH
SALT LAKE CITY

CLARENCE EDWARD DUTTON: AN APPRAISAL

I

THE MEN who made up the first generation of workers in the United States Geological Survey were in many ways a most remarkable group. They had, of course, the advantage that always lies with the pioneers of a movement—the whole western country, much of it unexplored, and most of it unsurveyed, lay before them inviting study. But even so they were men of large calibre. Powell, King, Hayden, Gilbert, Holmes, are names to conjure with in the history of American geology. Moreover, the achievements of these men did not stop with geology. Powell, a great organizer, "extraordinarily fertile in ideas," was the father not only of the Geological Survey, but of the Ethnological Bureau and the Reclamation Service as well. [1] King, who died prematurely and failed to fulfill the promise of his personality and abilities, was yet called by his friend John Hay "the best and brightest man of his generation," and Henry Adams admired him above any man he had ever met. Hayden and Gilbert, perhaps the greatest geologists of the group, were not so versatile as their companions, but Holmes reached distinction as geologist, artist, paleontologist, and ethnologist, and Dutton, Major Powell's particular protege, combined geological knowledge with literary skill as no other scientist has ever succeeded in doing it.

The concrete achievements of these men are common knowledge, and need no reciting here, but the less tangible elements of their distinction, the personal characteristics that made them interesting as men even more than as intellects, have begun to be forgotten. The personalities of Powell and King, it is true, have been perpetuated by their friends and admirers, but the others of the group are now hardly more than names. And in particular Major Dutton, one of the richest personalities of his time, has been so completely lost to us that only the kindness of his son [2] enables us to reconstruct him as he was.

Any appraisal of Dutton must recognize at once the fact that in spite of the comparative obscurity in which he lived (especially in his later years) he was a very unusual man. "To the casual

---

[1] For a good and sympathetic summary of Major Powell's personality and achievements see W. H. Hobbs, "John Wesley Powell, 1834-1902." *Scientific Monthly* 39: 519-29. December, 1934.

[2] A long letter from C. E. Dutton, Jr., summarizing his father's character and tastes, is included as Appendix A in W. E. Stegner, *Clarence Edward Dutton, Geologist and Man of Letters*, Thesis, State University of Iowa, 1935.

acquaintance," his son writes, "he must have seemed much like any man-in-the-street, except, perhaps, for his erect carriage and powerful physique, and a look in his eye which suggested the habit of command." Photographs of him in later life show a strong and rather handsome face, with a broad brow and a firm chin partially hidden by a well-trimmed van dyke. As for the "habit of command," acquired in the army, it came close to making his field work difficult, to say the least, for according to F. S. Dellenbaugh, [3] Dutton took with him to the western surveys the distinction between officer and man that prevailed in the service, and as a result had one or two sharp arguments with his men. Foreseeing difficulties, Professor Thompson led him aside and pointed out that the half-wild rugged frontiersmen who composed his party were too independent to submit to army discipline. After this advice, Dutton like a wise man threw aside his artillery training and almost immediately was the best-liked and most respected man in camp. The two old-timers so far discovered who worked with Dutton in southern Utah and Arizona both spoke in the highest terms of his geniality and friendliness. [4]

Underneath his reserved and unassuming appearance, however, this "man-in-the-street" concealed a teeming mind and powers of mental action that are decidedly rare. His reading and interests embraced many fields, and he could talk with ease and authority on all sorts of subjects. Although fond of company and good conversation, he preferred to listen unless specifically called upon, and then his contribution was generally sufficient to settle all argument. His conversational habits and his remarkable accuracy and variety of knowledge are well illustrated by a story told his son some years ago. The story came from a ship's captain who had had Dutton as a passenger between New York and Galveston. Part of the captain's duties consisted in entertaining his passengers, and he accordingly began a conversation with Dutton and a sugar planter from Louisiana on the first evening out. The planter waxed violent on tariffs, the captain sympathized and commented, and Dutton sat and smoked, until finally the captain asked Dutton's opinion. The result was a beautifully clear account of the whole sugar industry, foreign and domestic, so plainly and forcibly put that the planter retired. Night after night the captain, now with

---

[3] Letter to W. E. Stegner, February 18, 1935.

[4] One of these men, James Fennemore of Salt Lake City, was a photographer with Powell's second expedition in 1871-2, but did not go down the river with the party, joining them later at Kanab.

malice prepense, would start up an intellectual hare for Dutton to run, and on every occasion his passenger's knowledge, though modestly and almost reluctantly aired, amazed him. He at last gave up in despair: "It is a mystery to me how any man could ever learn and remember so much about so many subjects."

This many-sided intellectual interest, gained from wide reading (he called himself "omnibiblical"), would yet have been impossible without an extraordinarily retentive memory. When writing his books he never used notes, [5] but mulled his subject over mentally until all its parts fell into regular order. Then he wrote rapidly and from memory, and he practically never had to revise. On one occasion, when his son was acting as amanuensis, Dutton was compelled to get into shape immediately a long report embodying much tabular matter and statistical data. In one day he dictated over 11,000 words, and in the final check there were only two minor errors to be corrected. If his son failed to keep up in his typing, Dutton would repeat the whole sentence over, no matter how long, without hesitation and without change. Throughout his life he seems to have done his thinking afoot rather than seated, and the clear, pointed sentences emerged, as it were, from a moving cloud of cigar smoke. [6] The powerful memory that was one of Major Dutton's greatest intellectual weapons served for other things besides dictation of reports and papers, for in his last days he astonished his wife by quoting long passages of poetry remembered from his youth, poetry which she had never heard him mention before that time. In this trait he strongly resembled Macaulay, who was incidentally one of his literary favorites.

Along with his memory Major Dutton had another, and even rarer, attribute which deserves mention. This was the almost lightning-like swiftness with which ideas took form in his mind, and the ease with which they found expression. His organizing ability enabled him not only to compose easily and well, but also to speak extempore with remarkable fluency. Once when a lecturer at the National Museum had failed to appear Dutton was called on the telephone and asked to substitute. In the time it took to ride from the Geological Survey offices to the Museum he composed a lecture on "The Future of the West" for which both audience and dailies gave him great applause. Frequently also he lectured on regular programs, or before the Washington Philosophical Society, and became well-known as a talker in Washington. Perhaps his most popular lecture was the one given after his return from Hawaii,

---

[5] *Dictionary of American Biography*, Vol. V, p. 555.

[6] Letter from C. E. Dutton, Jr.

on "The Hawaiian Islands and People,"—a talk in which he in-
dulged all his liking for bantering tone and humorous anecdote, and
sprinkled his sentences with the alliterative expressions of which
he was fond. One of these last was the comparison of the full-
blooded and buxom girls of Hawaii with their "pale pious and
pulmonary sisters of the effete East."

Perhaps nothing illustrates the orderliness and concentration
of Dutton's mind better than his love for chess. From early youth
he loved the game, and on occasion played as many as seven
blind-fold games simultaneously. Sometimes he would get interested
in a chess problem and sit up over it all night, to be interrupted by
the call to breakfast. [7] Although he was forced to give it up in the
'90's because it interfered with his work, the same concentration
which he gave to chess is evident all through his books, and nothing
interested him so much as a problem whose solution was clouded
in mystery. This explains, at least in part, his interest in the more
speculative branches of geology, in seismology and volcanology,
where his mind could exercise itself by going beyond the facts and
"digging down to the tap-root for causes." Beginning at the very
start of his geological career, this speculative temper evidenced itself first in his development of the theory of isostacy, and spread
out later to other fields. His last work, as a matter of fact, was
a paper on "Volcanoes and Radioactivity," in which he thought
to have found the answer to the question of volcanic heat. As he
himself admitted, [8] there was no problem that he would rather
have solved, and it occupied his mind off and on from 1875 to his
death in 1912.

For all his unusual mental equipment, his accuracy, his speculative temper, his powers of memory and organization, Major
Dutton was an eminently human individual. There is something
about his personality that reminds one of Mark Twain, whom he
loved. He relished a good joke hugely, and had the very rare
ability to enjoy a hearty laugh. Until his illness in the '90's he
smoked prodigiously and like Twain he enjoyed billiards and base-
ball. There is much of Mark Twain's gusto, too, in the anecdotes
with which *Hawaiian Volcanoes,* the most personal of his books,
is embellished. In one place, [9] for example, he tells of coming down

[7] *Ibid.*

[8] Letter from Dutton to J. S. Diller, quoted in Diller's biographical memoir,
Bulletin of the Seisomological Society of America, Vol. I, pp. 137-42. December,
1911.

[9] Dutton, *Hawaiian Volcanoes.* U. S. G. S. Annual Report 4, Powell, 1882-3.
p. 148.

off the lava fields with his train on Sunday, and of being almost mobbed by a group of half-naked villagers for driving burros on the sabbath. In fact, whenever he allows himself to show through his books, which is not often the case, we catch glimpses of a genial, sane, many-sided nature, a personality whom it would have been a pleasure to know. And the story of his life, while not so sensational as that of his friend Powell, is yet interesting and in many ways typical of the whole group of pioneers in the western surveys.

II

OF THE boy Dutton little information is available. [10] He was born in Wallingford, Connecticut, on May 15, 1841, the son of Samuel and Emily (Curtis) Dutton. Like many others of his generation, he was prepared for college early, and was ready for matriculation at thirteen. At the last moment, however, his parents thought him too young, and held him back. Entering Yale at the age of fifteen, he was anything but an outstanding student. His scholastic record shows a consistently average performance—in modern parlance, a C average—and at graduation he ranked eighty-second in a class of one hundred nine. Like Emerson, he was outshone at college by many a lesser man, but a look at the Yale curriculum of the day indicates the reason. The courses, largely prescribed, were almost wholly in the fields of comparative theology, political economy, and classical languages. Since Dutton later showed a keen aptitude for mathematics in connection with his seismological studies, it would be interesting to know what records he made in that subject, but unfortunately grades for specific courses are not preserved. At any rate, he must have found a steady diet of Euclid, Herodotus, Cicero, and the Greek Testament somewhat fatiguing. That seems the only possible explanation for his mediocre work at Yale.

In other ways besides scholarship, however, Dutton showed himself no ordinary schoolboy. His principal triumph at college, aside from his gymnastic and rowing achievements (he liked to remember in later years that he rowed number 7 in Yale's first race against Harvard) was the winning of the Yale Literary Prize in 1859 with an essay on Charles Kingsley, the novelist. This prize essay, though it shows some youthfulness of thought and a religiosity that Dutton later lost, shows also wide reading, compact and close organization, and accurate and flexible phrasing. Its whole conception is unusual for a youth of eighteen—especially a C student. Already the prose style that gave life and color to his later writing was practically developed.

---

[10] A number of short sketches of Dutton's life have been published, besides the ordinary summaries in *Dictionary of American Biography* and *Who's Who in America*. Of these, the longest and most complete is contained in *The Biographical Record—Class of Sixty*, Boston, 1906. pp. 95-100. Other sources from which information has been taken for this paper are Diller's sketches (see note 8, above); G. P. Merrill, *First Hundred Years of American Geology*, New Haven, 1924; *Obituary Record of the Graduates of Yale University*, New Haven, 1915; an article in *American Journal of Science*, 4th Series, Vol. 33, pp. 387-8; and the administrative reports of Powell and King.

Although Dutton was in a very hot-bed of scientists, led by Dana, Brush, and Silliman, and was attending a school that produced King and Brewer, to mention only two of many great geologists who had their training at Yale, [11] his scientific studies were not to begin until after the Civil War. At school, his son tells us, he read widely in literature, history, and religion, and intended, urged by his parents, to enter the ministry. His first year in the Yale Theological Seminary, however, was interrupted by the war. Glad of the chance to avoid theology, [12] Dutton immediately enlisted, being appointed adjutant of the Twenty-First Connecticut Infantry.

His war career is not necessarily important here, as it is treated rather fully in *The Biographical Record—Class of Sixty*. It is only necessary to note that as a volunteer he saw hard and active service, participating in the battles of Fredericksburg. Harper's Ferry, Suffolk, and Harpeth River, and that he was severely wounded at Fredericksburg. More important is the fact that in December, 1863, he definitely gave up his pretensions to the ministry by passing an examination for a commission in the regular army, and was commissioned second lieutenant of Ordnance a month later. His choice of the Ordnance Corps was, according to his son, dictated by his love for mathematics. Three months after his commission he was married to Emeline C. Babcock, of New Haven, a lady of whom we know nothing except that she had a fine talent for the piano which Dutton appreciated and enjoyed greatly. From then on until the end of the war Dutton was largely engaged in arming and disarming troops, and except for a brief period with General Schofield, during which he participated in the bloody battle of Harpeth River, he saw no more action. The end of hostilities found him at the ordnance depot of the Army of the Potomac, and after the depot was closed he was stationed at the Watervliet Arsenal, West Troy, New York.

The assignment to West Troy was a happy accident. There were the great mills of the Bessemer Steel Works, and down the river at Albany was the paleontological Museum under James Hall and R. P. Whitfield. With characteristic earnestness Dutton devoted all his leisure time to scientific studies, spending much of the week in the steel works with Alexander Holley, and devoting Saturdays and Sundays to the study of paleontology in Albany. His scientific career dates from this five-year apprentice period under

---

[11] Dutton's brother, Colonel Arthur Henry Dutton, was a student at the Sheffield Scientific School for two years before entering West Point.

[12] Dutton's son tells us that he had no particular desire to enter the ministry, and protested that he "was too young to know his own mind." Certainly he had little talent for the profession, for in his later years he became completely agnostic.

noted masters. "On the Chemistry of the Bessemer Process," his first scientific paper, was read before the American Association for the Advancement of Science in 1869, and this beginning might have led Dutton on to further chemical research, in which he was highly interested, had he not been transferred again, this time to Frankford Arsenal in Philadelphia. During the year he was there he continued his two-fold studies, reading several papers before the Franklin Institute, the American Philosophical Society, and the Academy of Sciences.

His direct connection with geology did not come about until 1871, when his orders took him to the Washington Arsenal. Here his scientific interest was further aroused and his ambition stimulated by the distinguished members of the Washington Philosophical Society with whom he came in contact. Baird and Henry of the Smithsonian, Hilgard of the Coast Survey, Newcomb, Hall, and Harkness of the Naval Observatory, Hayden and Powell of the surveys of the West, made up a group from whom Dutton could learn much.

Out of his friendship with Hayden and Powell grew his keen interest in geology, in the study of which he spent two energetic years. At the end of that time, despite his lack of acquaintance with field work, he was sufficiently eminent as a structural geologist to receive an invitation from Powell to work on the survey of the Rocky Mountain Region. Powell, a keen and discriminating judge of men, saw the great possibilities of the young officer, and exerted every effort to bring Dutton into the circle of his "boys." It was not until 1875, however, that he and Professor Henry prevailed upon the War Department to detail Dutton to special duty with the survey, and in May, 1875, he went west with Powell's party to study the high plateaus of Utah, then largely virgin wilderness. He had in the beginning modestly protested his unfitness for the task, but Powell insisted, and the results more than justified his faith in his protege.

The monographs and reports which resulted from Dutton's twelve years of study in the West are part of that great series to which Powell, Hayden, Gilbert and Holmes all contributed, and which laid a broad base for the later geology of the western country. More than this, Dutton's works, as we shall see later, are probably the finest things from the standpoint of literary merit in the whole history of American geology. The science of geology, having got past its formative stage, has become more specialized, and it is probably true, as an eminent geologist told me recently, that a monograph written as Dutton's are written would never be accepted for publication by the Geological Survey today. But one must still be very glad that Dutton wrote as he did, for though his geology may be superseded to a certain extent, his literary merit is as great now as it ever was. *The High Plateaus of Southern Utah, The Physical Geology of the Grand Canyon District, The Tertiary History of the Grand Canyon, and Mount Taylor and the Zuñi Plateau* are, as Dellenbaugh has said, [13] unique for their combination of scientific accuracy with literary charm. *Hawaiian Volcanoes*, although outside the field of the western surveys, is fully as lucid and attractive, and has the additional distinction of being more personal in tone. Dutton's Hawaiian trip, taken in 1882 because he wished to study volcanoes in action before undertaking the study of the lava fields of Oregon, was really a vacation, and the report benefits by a lighter and more anecdotal style. The work on the Oregon volcanic fields was never completed by Dutton, since the Charleston earthquake called him back in 1886, and by the time his monograph on that catastrophe had been completed the western irrigation surveys had been initiated, and the directorship of the field work forced upon him.

The history of the early surveys which were in 1902 to be rejuvenated as the Reclamation Service is a rather dismal one. According to the sketch of Dutton in *The Biographical Record—Class of Sixty* (a sketch presumably furnished by Dutton himself, or by members of his family) Powell must have been somewhat to blame in his manipulation of Irrigation Survey funds. He apparently devoted part of these funds, for which he and Dutton had been instrumental in securing the appropriation, to the strengthening of the topographical work of the Geological Survey. Despite Dutton's warning that the procedure was not only ill-advised but illegal, Powell persisted, with the result that after the second year he was attacked by a group of western congressmen, an investigation was ordered, and the appropriation was discontinued. The "damaging report" after the investigation had a good deal to do with Powell's resignation as head of the Survey, and it had also a profound effect on Dutton's career.

After he had collected a group of hydrographers and had put them in training camp for the winter, he mapped his campaign and directed the stream-measurements and survey of the arid lands for one year. Then, when the appropriation was stopped, he might have gone on with his Oregon field work, but by now, tired of controversy and feeling that the Survey was no longer the same with Powell gone, he returned to active army duty in the autumn of 1890. Yet he did not entirely drop geological work. On a visit

---

[13] *Cambridge History of American Literature*, Vol. III, p. 159.

to Central America in 1891, .e made a report for the Nicaragua Canal Corporation, which was submitted to the congessional committee. Later in the same year, after ar infriendly Ordnance Corps commander had shunted him off to San ntonio, he was asked to make a report on the volcanic activities and the possibilities of earthquakes in the canal zone. Both reports w ordered printed by Congress. [14]

Banished to San Antonio, Major Dutton found outlet for his scientific propensities by visiting frequently the volcanic regions of Mexico. He may also have studied the region near San Antonio, where he might have found "the famous Balcones fault, the volcanic plug and neck near Uvalde, and Cretaceous and Tertiary formations, all of about the same age and character as the volcanic and diastrophic structures and the sediments of the Grand Canyon region." [15] But if he interested himself in these things, there is no record of it. In 1899 he was ordered back to Washington to an honorable and important post as assistant to the head of the Ordnance Corps, but finding the work unpleasant, and himself in poor health, he was retired at his own request in 1901.

Even then, in his sixties, and broken by ill health, he did not entirely give up his scientific pursuits, but in 1904 produced his important volume *Earthquakes in the Light of the New Seismology*, and also produced from time to time short articles on seismology and volcanology. In 1910 appeared the stimulating pamphlet "Volcanoes and Radioactivity," his last work. He died on January 4, 1912, at the home of his son in Englewood, New Jersey, where he had been living since his retirement.

---

[14] *Report on the Nicaragua Canal.* Sen. Mis. Doc. 97, 52nd Congress, 1st Sess. Ser. 2904, 1891-2. Vol. 2; also in House Rep. No. 2126, 54th Cong., 1st Sess., 1895-6. Vol. 9. *The Possible Effects of Earthquake Shocks on the Structures of the Canal.* Sen. Doc. 357, 57th Cong., 1st Sess. Ser. 4245, 1901-2. Vol. 26 pp. 55-62.

[15] This information as to the geological points of interest near San Antonio was kindly furnished me by Dr. A. C. Trowbridge, State Geologist of Iowa.

III

I HAVE said that Major Dutton combined geological knowledge and literary skill as few scientific men have. But as
a literary man he has never received anything like the
attention he merits. This is not to say that his literary
side was anything more than incidental, and it is not meant to
obscure the fact that his literary work, being all in the field of
nature description, belongs necessarily to a minor branch of literature. But within that field he can legitimately challenge any writers
of his time, and descriptions of the Grand Canyon and its surrounding region since his day have to a surprising extent been based on
what Dutton saw and how he saw it.

Let us admit in the beginning that even as a nature writer he
is limited. He has none of the all-embracing, all-devouring curiosity
of a Muir or Burroughs. He is, first and foremost, a geologist, and
his whole attention is fixed on the geological meaning of the regions
he traverses. Yet we must be cautious in applying this point too
stringently, because in the Plateau Country where Dutton did his
finest work there is little beside rocks. It is impossible to see this
country without thinking in almost exclusively geological terms, as
John Burroughs, who was stricken with an acute attack of "geological
fever" on his trip west, attests.

Even after all these concessions have been made, there is still
one popular misconception that must be removed before we may
consider Dutton's literary claims justly. It is the idea that no scientist
can also be a poet. Ruskin denied it; so did Emerson: "Empirical
science is apt to cloud the sight, and by the very knowledge of func-
tions and processes to bereave the student of the manly contemplation of the whole. The savant becomes unpoetic." [16] The more we
study Dutton's writing the more is evident the fundamental error at
the bottom of this statement, for a large part of his descriptive ex-
cellence comes about not in spite of his accurate scientific mind, but
because of it. Fine as are his panoramas and landscapes, it is in his
revelation of the scientific *method* as applied to nature writing that
his greatest value lies.

The first qualification of good description—knowledge of the
thing to be described—can in Dutton's case be dismissed in a sentence. He knew the canyon and the region surrounding it as well
as any man of his time, not excepting Major Powell, and he knew

[16] Emerson, *Works*, University Edition, Vol. 1, p. 70.

not only its appearance, but its meaning, which he derived from ser-
ious study of the climate, the stratification, the fossils, the fault lines,
the volcanic extravasations, the tilting of the beds, the drainage. And
instead of making his description ponderous and unwieldy with facts,
this accuracy of observation merely emphasizes and gives point to
the beauties of the country as they appear in Dutton's pages. As
an instance of how ignorance of geology can spoil description one
has only to turn to Joaquin Miller's fantastic notion that the Grand
Canyon was formed by the falling in of the crust over an under-
ground river [17]—a landslip that would involve several thousands of
cubic miles! And if this is not sufficient evidence that scientific accuracy is no hindrance to sound description, we may note the poem of
John Gould Fletcher, [18] in which the canyon is summarized in the
"strong invisible words," "It is finished." Dutton could never have
been guilty of so complete a misapprehension of everything that the
Grand Canyon means.

Dutton's writing is further helped by his scientific knowledge in
that nowhere is description allowed to stand alone. Literary landscape painting in itself is an inert and lifeless thing unless bolstered
by narrative or movement of one sort or another. But Dutton goes
one step further, and not only gives his landscapes vigor and life
by a narrative frame, but also adds a great deal to their stamina
by subordinating them to the larger purpose of geological interpretation. Throughout both *The High Plateaus* and *The Tertiary History*, the best of his books from the literary standpoint, the observer
thus manages to write sound science, and yet to enliven and beautify
it with gorgeous nature pictures. His eye is always on the object,
but he is capable of seeing it for its beauty as well as for its meaning.
In *The Tertiary History*, which contains as fine descriptive passages
as have ever been written by an American, this two-fold purpose results in a two-fold plan. Fearing that his field of study is too large for
easy comprehension, Dutton specifically states his intention of attack-
ing the reader through his imagination, so that one unfamiliar with
the country will not be left groping among a cold mass of facts. [19]
The product of his dual method is a book written in alternating
chapters of description and interpretation, with singularly happy
results.

---

[17] Joaquin Miller, "Grand Canyon of the Colorado." *Overland*, n. s., 37:786-90.
March, 1901.

[18] John Gould Fletcher, "The Grand Canyon of the Colorado." In *Breakers and
Granite*, New York, 1921. pp. 95-9.

[19] *Tertiary History*, pp. 5-6.

There is no doubt that as a writer Dutton aimed primarily at clearness, for he always treasured the remark made by Lord Kelvin to Archibald Geikie, who repeated it to him: "Would that more of our scientific writers were as lucid as Dutton." [20] But when he came in contact with the magnificent scenery of the Plateau Country his capacity for clearness outdid itself, and lucidity blossoms into stylistic beauty in page after page. His books are studded with grand panoramas, broad sweeps of vision that transmit the spirit of the country as no other descriptions do, and yet in *The High Plateaus,* in particular, we have a sense that he is repressing his enthusiasm almost too much. In *The Tertiary History,* on the other hand, we feel that he is no longer trying to keep his enthusiasm completely under, that he has frankly admitted that he cannot do his scientific job without allowing his appreciation at least equal space. Description as a consequence occupies more than half the book.

But Dutton's enthusiasm was not the wild ecstatic thing that most tourists bring to the contemplation of scenery. In the midst of colossal and inspiring scenes, he was yet able to keep his enthusiasm modulated, to graduate and scale the emotional power of landscape, and he avoided the treacherous zeal that describes everything in superlatives and leaves no words for the grandest scenes of all. He consciously refuses to describe the Vermillion Cliffs in too glowing terms, because adjectives, "weak and vapid enough at best," must be saved for the incomparable spectacle of the canyon. Knowing that in this country superlatives are dangerously easy, he guides himself by Emerson's rule: "The wise man shows his wisdom in separation, in gradation, and his scale of creatures and of merits is as wide as nature." [21]

Two more things indispensable to the nature writer Dutton possessed to a remarkable degree: an eye for color and an eye for form. These attributes, qualities of the artist or poet rather than of the scientist as he is ordinarily defined, were intimately blended in him with the cool observation, the meticulous accuracy, and the search for meanings typical of the scientist. The result of the combination makes us wish more scientists would study the arts, or that more artists were scientists.

Because the forms of the Plateau Country—mesas, buttes, plateaus, cliffs,—are so utterly different from the ordinary mountain scenery, it is obvious that they demand an altogether different terminology of description. This terminology Dutton borrowed from architecture, and with telling effect. Things are described in terms

---

[20] Letter from C. E. Dutton, Jr.

[21] Emerson, *Works,* University Edition, Vol. I, p. 44.

of Mansard roofs, wing-walls, pilasters; promontories have gable-
ends; rows of needles on the summit of a wall become lines of statu-
ary above a noble facade. Niches in the wall are described as being
overhung by arched lintels with spandrels. Since it is the author's
method to allow the scenery to unfold before the reader as it does
before the traveler, and since his observation is so definite and acute,
this persistent analogy with architectural forms results in singularly
pictorial effects. We are journeying, for example, down the Toro-
weap toward the Grand Canyon:

> Our attention is strongly attracted by the wall upon the eastern
> side. Steadily it increases its mass and proportions. Soon it
> becomes evident that its profile is remarkably constant. We
> did not notice this at first, for we saw in the upper valley only
> the summit of the palisade; but as the valley cuts deeper in the
> earth the plan and system begin to unfold. At the summit is
> a vertical ledge, next beneath a long Mansard slope, then a
> broad plinth, and last, and greater than all, a long, sweeping
> curve, gradually descending to the plain below. Just opposite
> to us the pediments seem half buried, or rather half risen out
> of the valley alluvium. But beyond they rise higher and higher
> until in the far distance the profile is complete. In this escarp-
> ment are excavated alcoves with openings a mile wide. As
> soon as we reach the first one, new features appear. The upper
> ledge suddenly breaks out into a wealth of pinnacles and stat-
> ues standing in thick ranks. They must be from 100 to 250
> feet high, but now the height of the wall is more than a thou-
> sand feet, and they do not seem colossal. Indeed, they look
> like a mere band of intricate fretwork—a line of balustrade
> on the summit of a noble facade. Between the alcoves the
> projecting pediments present gable-ends toward the valley
> plain. Yet whithersoever the curtain wall extends the same
> profile greets the eyes. The architect has adhered to his de-
> sign as consistently and persistently as the builders of the
> Thebaid or of the Acropolis. [22]

In passages such as this (and they are frequent) Dutton exhibits
a more than ordinary sensitiveness to form, line, and proportion.
He is not, however, content to stop with mere appreciation of natural
forms. Consistently he goes beyond the external appearances. Accustomed to search for geological meanings, he probably found it
natural to seek aesthetic meanings as well, and though he admits
that such speculations are matter for the metaphysician rather than
for the geologist, his contributions to the study of both form and
color are highly significant. He is perpetually amazed at the virtu-

---

[22] *Tertiary History,* pp. 84-5.

osity of nature, at the multiplicity of her devices for materializing beauty. Here, in this lonely desert, are forms which, "if planted upon the plains of central Europe, would have influenced modern art as profoundly as Fujiyama has influenced the decorative art of Japan." [23] Yet, he notes, they are altogether different from what we are accustomed to call beautiful, and in many cases violate practically all the rules laid down by generations of artists and aesthetes. Symmetry is rarely attained, yet its effects are produced in many other ways. In the Valley of the Virgin the walls are fretted and filagreed with ornate and suggestive decoration which, though not symmetrical, is yet beautifully consistent with the walls and towers which it adorns. One of Dutton's aesthetic principles, therefore, is that symmetry is not necessary, that it is "only one of an infinite range of devices" by which beauty may be achieved. [24]

The same insistence that nature should be viewed without preconceived notions of what is "correct" in scenery is apparent in Dutton's discussion of the colors of the Plateau Country. These too are unusual, unheard-of, altogether out of place in our formulas of beauty:

> The lover of nature, whose perceptions had been trained in the Alps, in Italy, Germany, or New England, in the Appalachians or Cordilleras, in Scotland or Colorado, would enter this strange region with a shock, and dwell there for a time with a sense of oppression, and perhaps with horror. Whatsoever things he had learned to regard as beautiful and noble he would seldom or never see, and whatsoever he might see would appear to him as anything but beautiful and noble. Whatsoever might be bold and striking would at first seem only grotesque. The colors would be the very ones he had learned to shun as tawdry and bizarre. The tones and shades, modest and tender, subdued yet rich, in which his fancy had always taken special delight, would be the ones which are conspicuously absent. But time would bring a gradual change. Some day he would suddenly become conscious that outlines which at first seemed harsh and trivial have grace and meaning; that forms which seemed grotesque are full of dignity; that magnitudes which had added enormity to coarseness have become replete with strength and even majesty; that colors which had been esteemed unrefined, immodest, and glaring, are as expressive, tender, changeful, and capacious of effects as any others. Great innovations, whether in art or literature,

---

[23] *Ibid.*, p. 150.

[24] *Ibid.*, p. 59.

in science or in nature, seldom take the world by storm. They must be understood before they can be estimated, and must be cultivated before they can be understood. [25]

It is unusual, at least, to find writing of this sort in a Gelogical Survey monograph. The justness of these perceptions and the value of the warning against hasty generalizations, in art or in nature, cannot be questioned, and the man who could express such ideas as clearly and beautifully as he has done deserves some sort of place in American literature.

But it is not for their aesthetic theories alone that Major Dutton's books have earned a place in literature. As has been mentioned, there are passages of supreme descriptive beauty, especially in *The Tertiary History*, and it is on one of these passages, picked from among many such, that it is best to close this paper. We are watching the transformation of the canyon through its many phases of light and shadow, as the sun draws toward its setting:

Throughout the afternoon the prospect has been gradually growing clearer. The haze has relaxed its steely glare and has changed to a veil of transparent blue. Slowly the myriads of details have come out, and the walls are flecked with lines of minute tracery, forming a diaper of light and shade. Stronger and sharper becomes the relief of each projection. The promontories come forth from the opposite wall. The sinuous lines of stratification, which once seemed meaningless, distorted, and even chaotic, now range themselves into a true perspective of graceful curves, threading the scallop edges of the strata. The colossal buttes expand in every dimension. Their long, narrow wings, which once were folded together and flattened against each other, open out, disclosing between them vast alcoves illumined with Rembrandt lights tinged with the pale refined blue of the ever-present haze. A thousand forms, hitherto unseen or obscure, start up within the abyss, and stand forth in strength and animation. All things seem to grow in beauty, power, and dimensions. What was grand before has become majestic, the majestic becomes sublime, and ever expanding and developing, the sublime passes beyond the reach of our faculties and becomes transcendent.

The colors have come back. Inherently rich and strong, though not superlative under ordinary lights, they now begin to display an adventitious brilliancy. The western sky is all aflame. The scattered banks of clouds and wavy cirrhus have caught the waning splendor, and shine with orange and crimson. Broad slant beams of yellow light, shot through the glory-

---

[25] *Ibid.,* pp. 141-2.

rifts, fall on turret and tower, on pinnacled crest and winding ledge, suffusing them with a radiance less fulsome, but akin to that which flames in the western clouds. The summit band is a brilliant yellow; the next below is pale rose. But the broad expanse within is a deep, luminous, resplendent red. The climax has now come. The blaze of sunlight poured over an illimitable surface of glowing red is flung back into the gulf, and, commingling with the blue haze, turns it into a sea of purple of most imperial hue—so rich, so strong, so pure that it makes the heart ache and the throat tighten. However vast the magnitudes, however majestic the forms, or sumptuous the decoration, it is in these kingly colors that the highest glory of the Grand Cañon is revealed. [26]

Description of this calibre is lamentably rare in the literary history of the Grand Canyon. Since the report of Lieutenant Ives there have been literally hundreds of pen-pictures of the canyon, not ten of which are worth the paper they are printed on. Even Muir and Burroughs, those masters of nature description, found themselves lost and groping for words when they tried to picture it, and it is significant to note that when they needed a particularly apt image they both on occasion turned to the pages of Dutton. Both give him the highest praise, and well they might, for a critical comparison of their essays with anyone of a half-dozen chapters in *The Tertiary History* shows the advantage, if anywhere, on Dutton's side. He was not a greater nature writer than Muir or Burroughs, since his work is too limited in its scope for such a claim, but within his range he wrote as well as any American has ever done. As literary interpretations of the geological aspects of natural scenery his works are as much a part of American literature as the works of Audubon or Agassiz are in their fields, and it will be a long day before anyone describes the Grand Canyon as well as Dutton described it in 1882.

---

[26] *Ibid.*, p. 155.

Bibliography of the Writings of
Clarence Edward Dutton

1. "Charles Kingsley, the Novelist." *Yale Literary Magazine*, XXV:101-6. Dec., 1859.

2. "The Chemistry of the Bessemer Process." *Proceedings of the American Association for the Advancement of Science*, 1869.

3. "The Causes of Regional Elevations and Subsidences." *Proceedings of the American Philosophical Society*, Vol. XII, pp. 70-2. 1871.

4. "A Criticism of the Contractional Hypothesis." *American Journal of Science*, 3rd Series, Vol. VIII, pp. 113-123. 1874.

5. "Critical Observations on Theories of the Earth's Physical Evolution." *Penn. Monthly*, Vol. VII, pp. 364-78, 417-31. May and June, 1876. Also in *Geological Magazine*, 2nd Series, Vol. III, pp. 322-28, 370-76. 1876. Abstract in *American Journal of Science*, 3rd Series, Vol. XII, pp. 142-5.

6. "Report on the Lithologic Characters of the Henry Mountain Intrusives." In Gilbert, G. K., *Report on the Geology of the Henry Mountains* (U. S. Geographical and Geological Survey of the Rocky Mountain Region): 61-65. 1877. Second edition, with title "The Intrusive Rocks of the Henry Mountains," 147-51. 1880.

7. "Irrigable Lands of the Valley of the Sevier River." In Powell, J. W., *Report on the Lands of the Arid Region of the United States*, 128-49. 1878.

8. "The Geological History of the Colorado River and Plateaus." *Nature*, Vol. XIX, pp. 247, 272. 1879.

9. *Report on the Geology of the High Plateaus of Utah.* U. S. Geographical and Geological Survey of the Rocky Mountain Region (Powell): XXXII, 1880. (Reviewed by J. D. Dana, *American Journal of Science*, 3rd series, Vol. XX, pp. 63-9. 1880.)

10. "The Causes of Glacial Climate" (with discussion). *Bulletin of the Philosophical Society of Washington.* Vol. II, pp. 43-8. 1880.

11. "On the Permian Formation of North America." *Bulletin of the Philosophical Society of America*, Vol. III, pp. 67-8. 1880.

12. "The Excavation of the Grand Canyon of the Colorado River" (abstract). *Proceedings of the American Association for the Advancement of Science*, Vol. XXX, pp. 128-30; *Science* (ed. Michels), Vol. II, pp. 453-4. 1881.

13. "On the Cause of the Arid Climate of the Western Portion of the United States" (abstract). *Proceedings of the American Association for the Advancement of Science*, Vol. XXX, pp. 125-8. 1882.

14. "The Physical Geology of the Grand Canyon District." United States Geological Survey, *Annual Report* II, pp. 47-166. 1882.

15. *The Tertiary History of the Grand Canyon District.* United States Geological Survey, Monograph II: xiv, 1882. (Reviewed by J. D. Dana, *American Journal of Science,* 3rd series, Vol. XXIV, pp. 81-9. 1882; also by Archibald Geikie, *Nature,* Feb. 15, 1883, p. 357.)

16. "Petrographic Notes on the Volcanic Rocks of the Yellowstone Park." United States Geographical and Geological Survey of the Territories (Hayden), *Annual Report* XII, part 2, pp.. 57-62. 1883.

17. "Recent Exploration of the Volcanic Phenomena of the Hawaiian Islands." *American Journal of Science,* 3rd series, Vol. XXV, pp. 219-26. 1883.

18. "The Hawaiian Islands and People." A lecture delivered at the National Museum, Feb. 9, 1884. (Separately published.) Washington, 1884.

19. *Hawaiian Volcanoes.* United States Geological Survey, *Annual Report* IV, pp. 75-219. 1884.

20. "The Geology of the Hawaiian Islands." *Bulletin of the Philosophical Society of Washington,* Vol. VI, pp. 13-14. 1884.

21. "The Volcanic Problem Stated." *Bulletin of the Philosophical Society of Washington,* Vol. VI, pp. 87-92. 1884.

22. "The Effect of a Warmer Climate upon Glaciers." *American Journal of Science,* 3rd series, Vol. XXVII, pp. 1-18. 1884.

23. "The Basalt Fields of New Mexico." *Nature,* Vol. XXXI, pp. 88-9. 1884. Abstract in *American Naturalist,* Vol. XIX, pp. 390-91. 1885.

24. "Mount Taylor and the Zuni Plateau." United States Geological Survey, *Annual Report* VI, pp. 105-198. 1885.

25. "The Volcanoes and Lava Fields of New Mexico." (Abstract, with discussion by J. W. Powell.) *Bulletin of the Philosophical Society of Washington,* Vol. VII, pp. 76-79. 1885.

26. "The Latest Volcanic Eruption in the United States" (Lassen Peak, 1883). *Science,* Vol. VI, pp. 46-7. 1885.

27. "Crater Lake, Oregon, a Proposed National Reservation." *Science,* Vol. VII, pp. 179-82. 1886.

28. "The Submerged Trees of the Columbia River." *Science,* Vol. IX, pp. 82-4. 1886.

29. (With Hayden, Everett) "Abstract of the Results of the Investigation of the Charleston Earthquake." *Science,* Vol. IX, pp. 489-501. 1887.

30. "The Charleston Earthquake." *Science,* Vol. X, pp. 10-11, 35-6. 1887.

31. "On the geologic nomenclature in general and the classification nomenclature and distinctive characteristics of the pre-Cambrian formation and the origin of serpentine." *International Congress of Geology, American Committee Reports,* 1888: A, 1888, pp. 71-3.

32. "On the Depth of Earthquake Foci" (abstract). *Bulletin of the Philosophical Society of Washington,* Vol. X, pp. 17-19. 1888.

33. (With Newcomb, Simon) "The Speed of Propagation of the Charleston Earthquake." *American Journal of Science*, 3rd series, Vol. XXXV, pp. 1-15, 1888.

34. *The Charleston Earthquake of August 31, 1886.* United States Geological Survey, *Annual Report* IX, pp. 203-528. 1889.

35. "On Some of the Greater Problems of Physical Geology" (with discussion by G. K. Gilbert and R. S. Woodward). *Bulletin of the Philosophical Society of Washington*, Vol. XI, pp. 51-64, 536-37. 1889.

36. "Atlantic and Pacific Railroad." *Macfarlane's Geo. Railway Guide*, 2nd edition, 1890, p. 323.

37. "The Crystalline Rocks of Northern California and Southern Oregon.' *International Geological Congress*, IV, London, 1888; *C. R.*, 176-79. 1891.

38. "Volcanoes and Earthquakes, Nicaragua and and Costa Rica." In *The Inter-Oceanic Canal of Nicaragua* (published by the Nicaragua Canal Construction Company): New York, 1891. pp. 73-78.

39. "A General Description of the Volcanic Phenomena Found in That Portion of Central America Traversed by the Nicaragua Canal.—The Possible Effects of Earthquake Shocks on the Structures of the Canal." *Senate Document 357*, 57th Congress, 1st session. Serial 4245. Vol. XXVI, pp. 55-62. 1901-2.

40. "Report on the Nicaragua Canal." *Senate Miscellaneous Documents*, No. 97. 52nd Congress, 1st session. Serial 2904. Vol. II. 1891-2. Also in *House Reports*, No. 2126. 54th Congress, 1st session, Vol. IX. 1895-6.

41. *Earthquakes in the Light of the New Seismology.* New York, 1904.

42. "Volcanoes and Radioactivity." (Read before the National Academy of Sciences, April 17, 1906.) Englewood, N. J., 1906. Also in *Journal of Geology*, Vol. XIV, pp. 259-68. 1906. Also in *Popular Science Monthly*, Vol. LXVIII, pp. 543-550. 1906.